How to Survive A Canadian Winter

by
Rebecca Broeders

AuthorHouse™
1663 Liberty Drive
Bloomington, IN 47403
www.authorhouse.com
Phone: 1-800-839-8640

© 2012 Rebecca Broeders. All Rights Reserved.

No part of this book may be reproduced, stored in a retrieval system,
or transmitted by any means without the written permission of the author.

Published by AuthorHouse 10/16/2012

ISBN: 978-1-4772-7781-2 (sc)

Any people depicted in stock imagery provided by Thinkstock are models,
and such images are being used for illustrative purposes only.
Certain stock imagery © Thinkstock.

This book is printed on acid-free paper.

Because of the dynamic nature of the Internet, any web addresses or links contained in this book may have changed
since publication and may no longer be valid. The views expressed in this work are solely those of the author and do not
necessarily reflect the views of the publisher, and the publisher hereby disclaims any responsibility for them.

Hello my name is Eli
And on me you must rely
To teach you all about winter's cold
A lesson that I was once told
About snow and ice and freezing air
About hats and mitts and long underwear

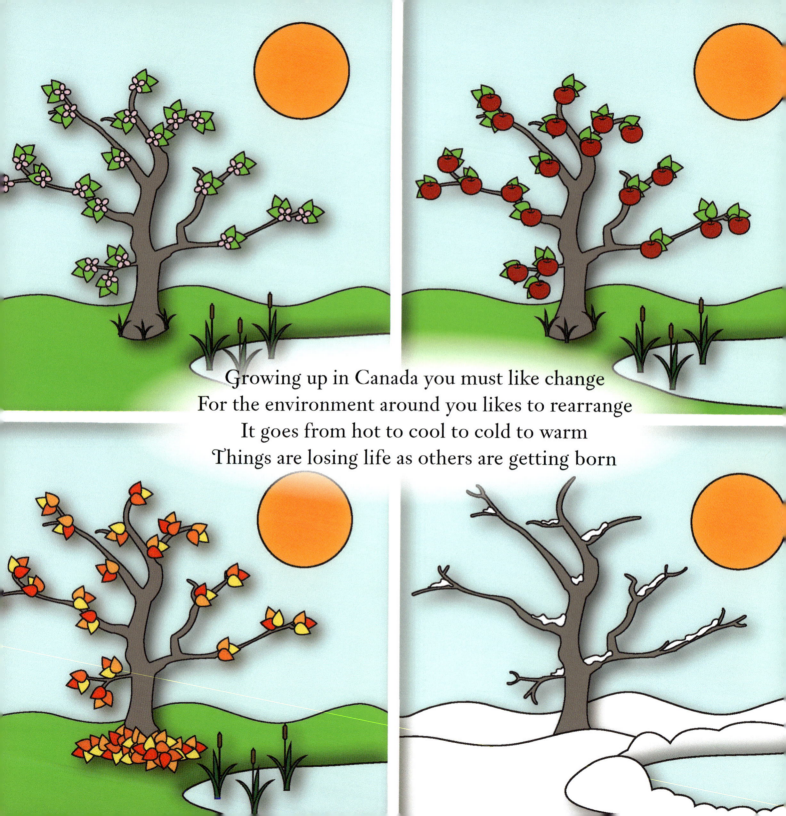

Growing up in Canada you must like change
For the environment around you likes to rearrange
It goes from hot to cool to cold to warm
Things are losing life as others are getting born

Winter is said to begin on December's twenty-first day
Though I think it begins when the snow's on its way
That magical day when the first snowfall arrives
Is always the best day of all Canadian kids' lives

And when that day comes at the end of each year
I can't stop thinking about all the fun drawing near
All the creating of forts, snowmen and our very own ice rink
All the kids in their snowsuits with their cheeks of rosy pink

My favourite winter activity is the sport of ice hockey
It's a really fun game that my daddy has taught me
Did you know that ice hockey is a Canadian creation?
For many years it's been played all over our nation

I also like to go snowshoeing and toboggan on snow mounds
It's fun to roll up parts of a snowman really big and round

I know all of these exciting activities are on everyone's mind
But have you ever thought about what makes weather of this kind?
Why is our winter cold? And what makes the ice freeze?
Why isn't the sun in the sky very long and more people sneeze?

Well, I can tell you why winter is cold by showing you in my atlas
And by teaching you a little about the tilt of the Earth's axis
It's tilted and turning and going around the sun non-stop
Canada has cold winters because it is on the planet's top

The cold weather has to do with the angle of the sun's rays
They are pointing at the Earth in a more slanted way
That's why there are longer nights and shorter days
That's why it's cold enough that the snow on the ground stays

That snow that's on the ground comes from the clouds above
It's made of the same thing that rain is made of
That's right! Water! Which also turns into ice
It's the cold weather that turns rain into something so nice

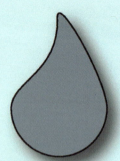

The water that evaporates for the cloud to take hold
Doesn't stay as liquid because the air is so cold
It turns into little crystals of ice and forms into snow
It's clusters of those crystals that fall when the clouds let go

When the air is that cold it can hurt our skin
So it's important what clothes you choose to play in
The number one thing you must remember is to bundle up tight
For that means more time outside if you do it right

All that fun is waiting for me outside my door
But am I ready to have fun and explore?
There are lots of things to remember on a snowy day
So many warm things to keep us safe while we play

If you start from top to bottom you'll have a better chance
Hat, scarf, coat, mittens, snow boots and snow pants
But before you get all dressed make sure you use the john
For it becomes quite a hassle after putting all that on

Going outside is fun and it's also good for you
All that fresh air will keep you good as new
Have you ever noticed in the winter more people get ill?
It's not because of the cold winter's chill

It's because they're staying inside more
And germs can spread when everyone's indoors
So bundle up from your feet to your head
And build and skate and ski and sled!

CPSIA information can be obtained
at www.ICGtesting.com
Printed in the USA
LVIC04n0752221016
509789LV00008B/19